Aims and Scope

The primary aim of *Advances in SOFC Technology: GDC-SDC Composite Cathodes* is to explore and analyze recent advancements in Solid Oxide Fuel Cells (SOFCs), focusing on the development and application of gadolinium-doped ceria (GDC) and samarium-doped ceria (SDC) as composite cathodes. This document seeks to provide a comprehensive understanding of the fundamental principles behind SOFC operation, emphasizing the role of GDC and SDC in enhancing electrochemical performance, catalytic efficiency, and long-term stability.

The scope includes the following areas:

- Fundamentals of SOFCs: Introducing SOFC technology, its key components, and operational principles, with a focus on electrochemical reactions, particularly the oxygen reduction reaction (ORR) occurring at the cathode.

- Composite Cathode Materials: Analyzing the properties of GDC and SDC, exploring their individual and combined advantages in composite cathodes, and detailing their roles in improving ionic conductivity, thermal stability, and overall cell performance.

- Fabrication Techniques: Reviewing advanced synthesis and fabrication methods—such as sol-gel, screen printing, and spray deposition—that impact the microstructure and efficiency of GDC-SDC composite cathodes, highlighting scalability for industrial use.

- Challenges and Innovations: Addressing technical challenges like interfacial compatibility, porosity control, and particle dispersion, alongside recent innovations in nanostructuring and scalable manufacturing processes that support high-performance, durable SOFCs.

This document is targeted at researchers, engineers, and clean energy professionals, offering insights and strategies for optimizing SOFCs through material science and advanced fabrication techniques, with the ultimate goal of enhancing the applicability and sustainability of fuel cell technology in modern energy systems.

Preface

Solid Oxide Fuel Cells (SOFCs) represent a promising technology for efficient and sustainable power generation, particularly as global energy demands grow and environmental impacts of traditional energy sources intensify. This document, Advances in SOFC Technology: GDC-SDC Composite Cathodes, provides an in-depth examination of the developments and innovations in SOFCs, with a focus on the application of gadolinium-doped ceria (GDC) and samarium-doped ceria (SDC) composite cathodes. By exploring the principles, material properties, and fabrication techniques associated with these advanced composites, this work aims to offer insights into the catalytic efficiency, operational stability, and material resilience that make GDC-SDC cathodes particularly valuable in SOFC applications. It covers both fundamental concepts of SOFC operation and the challenges encountered in their fabrication, including compatibility, microstructural control, and porosity management. Furthermore, the text highlights recent innovations and scalable techniques essential for moving SOFC technology from research to practical, commercial applications. This comprehensive study is intended to serve as a resource for researchers, engineers, and professionals in clean energy fields who are interested in advancing the capabilities and durability of SOFCs, particularly through the development of composite cathode materials.

Chapter 1: Fundamentals of Solid Oxide Fuel Cells and Composite Cathodes

Introduction to SOFCs

Basic working principles of SOFCs

Solid Oxide Fuel Cells (SOFCs) operate on the principle of electrochemical conversion of chemical energy into electrical energy, utilizing a solid oxide electrolyte to conduct oxygen ions from the cathode to the anode. The typical configuration includes a dense electrolyte sandwiched between porous anode and cathode materials, where oxidation of hydrocarbon fuels occurs at the anode and reduction of oxygen takes place at the cathode (Singhal, 2000; Ahmad, 2023; Ramasamy, 2024). SOFCs are characterized by their high energy conversion efficiency, fuel flexibility, and capability to operate at elevated temperatures (600-900°C) (Hajimolana et al., 2012; Zhou et al., 2023). This makes them suitable for various applications, particularly in stationary power generation and microgrid systems (Wu, 2023). Recent advancements in materials and cell design have further enhanced their performance, allowing for the direct utilization of diverse fuels, including hydrogen and ammonia (Vázquez et al., 2015; Feng et al., 2012). The integration of SOFCs with other systems, such as gas turbines, can also improve overall energy efficiency (Hajimolana et al., 2012).

Key components: Anode, cathode, electrolyte

Solid Oxide Fuel Cells (SOFCs) consist of three key components: the anode, cathode, and electrolyte, each playing a crucial role in the electrochemical process. The anode showed in figure 1, typically composed of nickel and yttria-stabilized zirconia (YSZ), facilitates the oxidation of fuel, allowing for the generation of electrons and oxygen ions (Nawaz et al., 2018; Abbas et al., 2013). The cathode, often made from perovskite materials, is responsible for the reduction of oxygen, where it combines with the incoming oxygen ions to form oxygen gas (Bi et al., 2018; Chen et al., 2016). Recent advancements in cathode materials have focused on enhancing catalytic activity and stability, which are critical for improving overall cell performance (Ahmad, 2023; Chen et al., 2016).

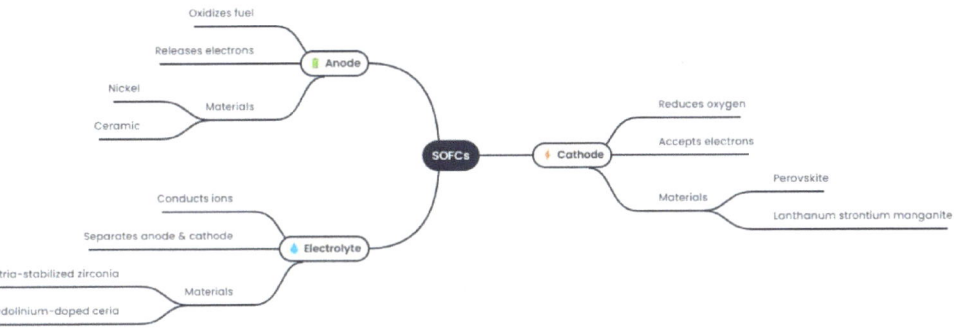

Figure 1. Components of SOFCs

The electrolyte, usually a dense ceramic material like YSZ, serves to conduct oxygen ions from the cathode to the anode while being impermeable to gases, thereby maintaining the separation of reactants (Gunawan et al., 2021; Zhu et al., 2011). The efficiency of SOFCs heavily relies on the ionic conductivity of the electrolyte and the interfacial interactions between the electrodes and the electrolyte (Zheng et al., 2016; Dong et al., 2013). Innovations in material science continue to enhance the performance and durability of these components, making SOFCs a promising technology for clean energy applications (Wang et al., 2022).

Role of Cathodes in SOFCs

Oxygen reduction reaction (ORR)

The oxygen reduction reaction (ORR) is a critical process in the operation of solid oxide fuel cells (SOFCs), occurring at the cathode where oxygen molecules are reduced to oxygen ions. This reaction involves several steps, including the adsorption of O_2, its dissociation, and the incorporation of oxygen ions into the cathode material Cao et al. (2019). The efficiency of the ORR is significantly influenced by the catalytic properties of the cathode materials, which are primarily mixed ionic and electronic conductors, often based on perovskite structures (Zhang et al., 2022; Nagasawa & Hanamura, 2017).

Recent studies have highlighted the importance of optimizing cathode compositions to enhance ORR kinetics, particularly at intermediate temperatures (600-800°C), where traditional materials may exhibit sluggish performance due to increased polarization resistance (Chen et al., 2016). Innovations such as the incorporation of nanoparticles and the development of composite materials have shown promise in improving catalytic activity and stability under operational conditions (Zhou et al., 2011; Choi et al., 2016). Furthermore, the design of cathodes that facilitate a high three-phase boundary (TPB) is essential for maximizing the active sites available for the ORR (Han et al., 2023). Overall, advancements in cathode materials and structures are crucial for enhancing the performance and durability of SOFCs.

Material requirements for cathodes (electronic conductivity, ionic conductivity, thermal stability)

The performance of cathodes in Solid Oxide Fuel Cells (SOFCs) is critically dependent on several material requirements, including electronic conductivity, ionic conductivity, and thermal stability. High electronic conductivity is essential for efficient electron transport during the oxygen reduction reaction (ORR), with values exceeding 100 S/cm being preferred to ensure effective catalytic activity across the cathode surface Jo et al. (2021). Materials such as lanthanum strontium cobalt iron oxide (LSCF) and perovskite-type oxides like $BaCoO_3$ and $SrCoO_3$ are commonly utilized due to their favorable electronic and ionic conductive properties (Cetin et al., 2014).

Ionic conductivity is also crucial, as it allows for the transport of oxygen ions from the cathode to the electrolyte. Mixed ionic-electronic conductors (MIECs) are often employed to facilitate this process, with materials like $La_{1-x}Sr_xCo_{1-x}Fe_xO_3-\delta$ showing promise (Jia et al., 2019). Additionally, the thermal stability of cathode materials is vital to withstand the high operating temperatures of SOFCs (typically 600-800°C) without degradation. The compatibility of the thermal expansion coefficients between the cathode and the electrolyte, such as yttria-stabilized zirconia (YSZ), is also essential to minimize mechanical stresses during operation (Cetin et al., 2014; Lin et al., 2017).

Recent advancements in cathode materials, including the development of composite structures and the incorporation of nanoparticles, aim to enhance these properties, thereby improving the overall efficiency and longevity of SOFCs (Ahmad, 2023).

Introduction to Composite Cathodes

Definition and types of composite cathodes

Composite cathodes in Solid Oxide Fuel Cells (SOFCs) are defined as materials that combine two or more components to enhance electrochemical performance, particularly for the oxygen reduction reaction (ORR). These cathodes can be categorized into several types based on their composition and structure. One common type is the mixed ionic-electronic conductor (MIEC) composite, which integrates materials that facilitate both ionic and electronic conduction, such as lanthanum strontium cobalt ferrite (LSCF) combined with gadolinium-doped ceria (GDC) Han et al. (2017). This combination improves overall conductivity and catalytic activity, leading to enhanced performance in SOFC applications.

Another type includes layered or fibrous composites, which emphasize the active surface area and promote efficient gas diffusion and reaction kinetics (Lee, 2023). These structures can significantly improve the ORR kinetics by increasing the three-phase boundary (TPB) where the gas, electrolyte, and electrode meet (Nagasawa & Hanamura, 2017). Additionally, composites that utilize carbonate materials have been explored for low-temperature SOFCs, where the presence of carbonate enhances porosity and facilitates ionic transport (Agun et al., 2015).

Furthermore, the development of core-shell structures in composite cathodes has emerged as a promising approach, providing improved thermal stability and electrochemical performance by optimizing the microstructure (Chen et al., 2014). Overall, the strategic design of composite cathodes plays a crucial role in advancing SOFC technology by enhancing efficiency, stability, and operational flexibility.

Benefits of composite materials in SOFCs

Composite materials in Solid Oxide Fuel Cells (SOFCs) offer several significant benefits that enhance their performance and operational efficiency as shown in figure 2. One of the primary advantages is the improved electrochemical activity resulting from the combination of different materials. For instance, the integration of lanthanum strontium cobalt ferrite (LSCF) with gadolinium-doped ceria (GDC) in composite cathodes has been shown to enhance both ionic and electronic conductivity, leading to better performance in the oxygen reduction reaction (ORR) Han et al. (2017). This synergy allows for a more effective charge transfer and reduces polarization losses, which are critical for overall cell efficiency.

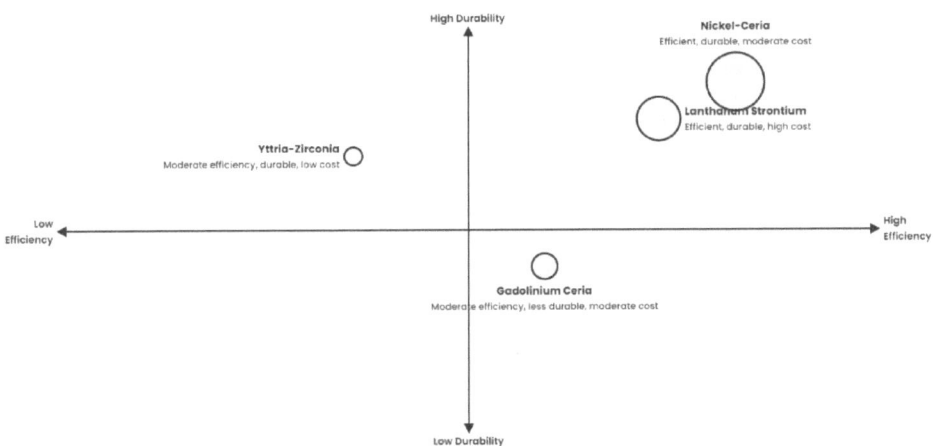

Figure 2: Benefits of Composite Materials in SOFCs

Additionally, composite materials can provide enhanced thermal stability and mechanical integrity. The combination of materials with complementary thermal expansion coefficients minimizes stress at the interfaces, thereby improving durability under operational conditions (Ko et al., 2012). For example, the LSCF-GDC composite cathodes exhibit greater resistance to thermal degradation compared to single-phase materials, which is essential for maintaining performance over extended periods (Ko et al., 2012).

Moreover, the use of composites allows for tailored porosity and microstructure, which can optimize gas diffusion and increase the active surface area for reactions (Agun et al., 2015). This is particularly beneficial in low-temperature SOFCs, where enhanced porosity can facilitate ion transport and improve overall performance (Agun et al., 2015). Overall, the strategic design of composite materials in SOFCs not only boosts efficiency but also contributes to the longevity and reliability of the fuel cells.

Overview of GDC and SDC Materials

Properties of Gadolinium-doped Ceria (GDC)

Gadolinium-doped ceria (GDC) is a prominent material used in solid oxide fuel cells (SOFCs) due to its excellent ionic conductivity and stability at elevated temperatures. One of the key properties of GDC that showed in figure 3 is its high ionic conductivity, which is significantly enhanced by the introduction of gadolinium as a dopant. This doping creates oxygen vacancies in the ceria lattice, facilitating the movement of oxygen ions, which is critical for the electrochemical reactions occurring in SOFCs Li et al. (2010)Alemayehu et al., 2022). Studies have shown that GDC exhibits one of the highest ionic conductivities among doped ceria materials, making it suitable for use as an electrolyte or as a component in composite cathodes (Navarrete et al., 2016; Alemayehu et al., 2022).

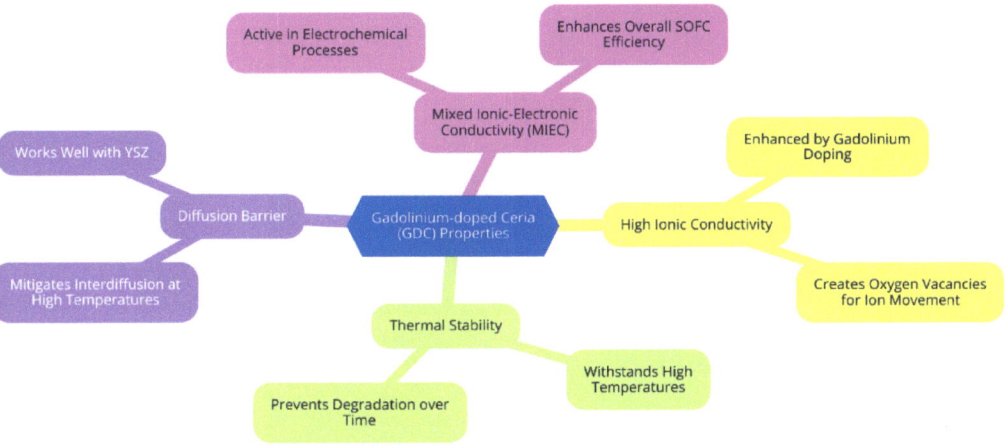

Figure 3. Properties of GDC

In addition to its ionic conductivity, GDC also demonstrates good thermal stability, which is essential for maintaining performance in the harsh operating conditions of SOFCs (Mulligan et al., 2022; Coppola et al., 2018). The material's ability to withstand high temperatures without significant degradation allows it to function effectively over extended periods. Furthermore, GDC can serve as a diffusion barrier layer between the cathode and electrolyte, mitigating interdiffusion issues that can arise at high temperatures, particularly when used in conjunction with yttria-stabilized zirconia (YSZ) (Coppola et al., 2018; Brito et al., 2011).

Moreover, GDC's mixed ionic and electronic conductivity (MIEC) properties enable it to participate actively in the electrochemical processes within the fuel cell, enhancing the overall efficiency of the system (Mulligan et al., 2022;

Futamura et al., 2017). This versatility makes GDC a valuable component in various configurations of SOFCs, contributing to improved performance and durability.

Properties of Samarium-doped Ceria (SDC)

Samarium-doped ceria (SDC) is a notable electrolyte material in solid oxide fuel cells (SOFCs) due to its superior ionic conductivity and thermal stability as shown in figure 4. One of the primary properties of SDC is its high ionic conductivity, which is significantly enhanced by the incorporation of samarium ions into the ceria lattice. This doping creates oxygen vacancies, facilitating the movement of oxygen ions, which is crucial for the electrochemical reactions in SOFCs Yang et al. (2016). Research indicates that SDC exhibits ionic conductivities comparable to or exceeding those of traditional electrolytes like yttria-stabilized zirconia (YSZ), particularly at lower operating temperatures (500-700°C) (Ali et al., 2013; Rahmanipour et al., 2017).

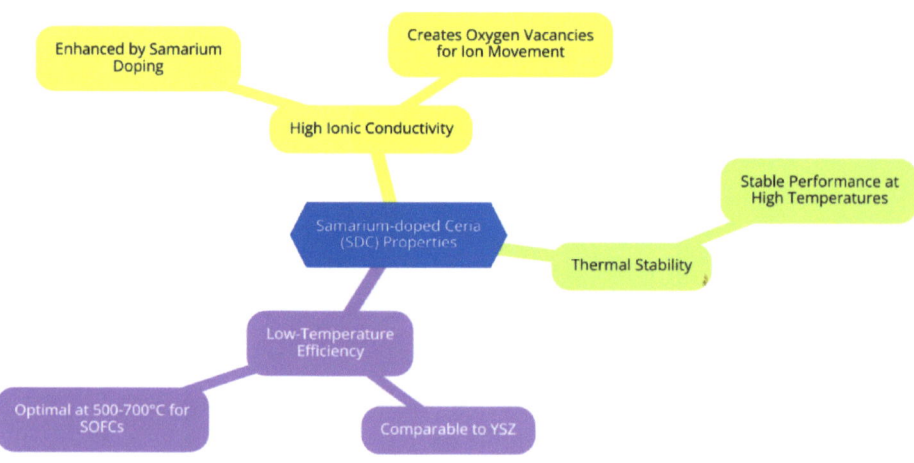

Figure 4. Properties of SDC

In addition to its ionic conductivity, SDC demonstrates good thermal stability, allowing it to maintain performance under the high-temperature conditions typical of SOFC operation (Deng et al., 2015). The material's stability is further enhanced when combined with other components, such as alkaline carbonates, which can improve its chemical compatibility and mechanical stability (Hoa et al., 2016; Ali et al., 2018).

Furthermore, SDC exhibits mixed ionic-electronic conductivity (MIEC) behavior, which can be advantageous in certain applications, although it may also lead to challenges in efficiency due to electronic conduction in reducing atmospheres (Ali et al., 2013; Rahmanipour et al., 2017). Overall, the unique properties of SDC make it a promising candidate for use in low-temperature SOFCs, contributing to improved performance and durability.

Synergistic effects of combining GDC and SDC in composite cathodes

The synergistic effects of combining gadolinium-doped ceria (GDC) and samarium-doped ceria (SDC) in composite cathodes for solid oxide fuel cells (SOFCs) can significantly enhance their electrochemical performance. One of the primary benefits of this combination is the improvement in ionic conductivity, as both GDC and SDC exhibit high ionic conductivities due to the presence of oxygen vacancies created by doping Ren et al. (2018). When used together, they can complement each other, resulting in a composite cathode that maintains high ionic transport while also benefiting from the unique properties of each material.

Moreover, the combination of GDC and SDC can lead to a reduction in polarization resistance (R_p), which is crucial for enhancing the overall efficiency of the cathode (Baharuddin et al., 2016). The presence of both materials can expand the reaction zone, allowing for more active sites for the oxygen reduction reaction (ORR) to occur, thus improving the cathode's catalytic activity (Kikuchi et al., 2011). This is particularly important in intermediate temperature SOFCs, where maintaining high performance at lower temperatures is essential for practical applications (Irshad et al., 2016).

Additionally, the mixed ionic-electronic conductivity (MIEC) of the composite can facilitate better charge transfer during operation, further enhancing performance (Li et al., 2016). The structural compatibility of GDC and SDC also contributes to the stability of the composite cathode, reducing the likelihood of phase segregation or degradation over time (Deng et al., 2015). Overall, the synergistic effects of combining GDC and SDC in composite cathodes can lead to improved performance, stability, and efficiency in SOFC applications.

Historical Development and Current Trends in GDC-SDC Cathode Research

The historical development and current trends in research on composite cathodes utilizing gadolinium-doped ceria (GDC) and samarium-doped ceria (SDC) reflect significant advancements in solid oxide fuel cell (SOFC) technology as shown in figure 5. Initially, the focus was on improving the ionic conductivity of electrolytes, with both GDC and SDC emerging as promising candidates due to their high ionic conductivities at intermediate temperatures Park et al. (2005)Ali et al., 2018). Research in the early 2000s highlighted the potential of these materials to enhance the performance of SOFCs, particularly in terms of reducing polarization resistance (Nie et al., 2010; Zhang et al., 2009).

As the understanding of the electrochemical properties of GDC and SDC evolved, studies began to explore their synergistic effects when combined in composite cathodes. This approach aimed to leverage the strengths of both materials, resulting in improved electrochemical activity and stability (Liu et al., 2013; Zhang et al., 2014). Recent studies have shown that the incorporation of GDC into SDC-based cathodes can significantly enhance the oxygen reduction reaction (ORR) kinetics, thereby improving overall cell performance (Takamura et al., 2023; Li et al., 2016).

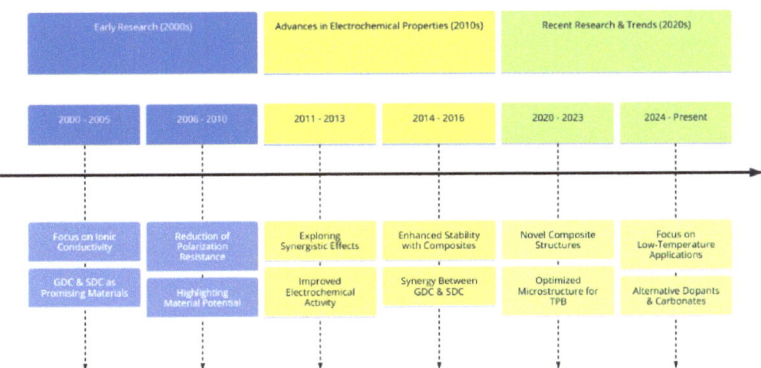

Figure 5. Historical Development and Trends in GDC-SDC Cathode Research

Current trends in research emphasize the development of novel composite structures and fabrication techniques to optimize the microstructure and enhance the three-phase boundary (TPB) for improved reaction kinetics (Sutradhar et al., 2011; Milcarek & Ahn, 2020). Additionally, there is a growing interest in the use of these composites in low-temperature applications, where they can provide efficient performance while minimizing degradation (Yue & Irvine, 2012; Rahmanipour et al., 2017). The exploration of alternative dopants and the integration of carbonate phases with GDC and SDC are also being investigated to further enhance ionic conductivity and electrochemical performance (Ali et al., 2018; Li et al., 2016). Overall, the ongoing research into GDC-SDC composite cathodes continues to push the boundaries of SOFC technology, aiming for higher efficiency and durability in practical applications.

Chapter 2: Synthesis and Fabrication Techniques of GDC-SDC Cathodes

Material Synthesis for GDC and SDC

Methods for preparing GDC and SDC powders (solid-state reactions, wet chemical methods, etc.)

The preparation of GDC (Gadolinium-Doped Ceria) and SDC (Samarium-Doped Ceria) powders can be achieved through various methods, including solid-state reactions and wet chemical techniques. Solid-state reactions often involve high-temperature calcination of mixed oxide precursors, which can lead to the formation of desired phases, as seen in the synthesis of SDC powders via co-precipitation and subsequent calcination at elevated temperatures (Mishina et al., 2020). This method allows for the control of stoichiometry and phase purity, essential for optimizing the electrochemical properties of the resulting materials (Min & Liu, 2013).

Wet chemical methods, such as the sol-gel process and glycine-nitrate process, have gained attention due to their ability to produce homogeneous and fine powders with controlled microstructures (Yang et al., 2010; , Zhang et al., 2015). For instance, the glycine-nitrate method has been shown to yield SDC powders with uniform particle size, enhancing the performance of fuel cell anodes (Yang et al., 2010). Additionally, the sol-gel method allows for the synthesis of SDC with high specific surface areas, which is beneficial for applications in solid oxide fuel cells (Myoujin et al., 2011). Overall, the choice of synthesis method significantly influences the microstructural and electrochemical properties of GDC and SDC materials.

Optimizing dopant concentration and particle size

Optimizing dopant concentration and particle size in Gadolinium-Doped Ceria (GDC) and Samarium-Doped Ceria (SDC) is crucial for enhancing their electrochemical performance in solid oxide fuel cells (SOFCs). Studies indicate that increasing the dopant concentration can improve ionic conductivity, but excessive doping may lead to phase instability and reduced performance (Fu et al., 2012). For instance, the optimal concentration of SDC dopant has been found to significantly influence the electrochemical properties, with a balance needed to avoid secondary phase formation (Fu et al., 2012).

Particle size also plays a critical role; smaller particles generally provide a larger surface area, enhancing the triple phase boundary necessary for electrochemical reactions (Ahmad et al., 2012). Research shows that milling speed affects particle size, where higher speeds yield finer powders that improve the electrochemical reaction sites (Ahmad et al., 2012). Additionally, controlling the sintering temperature is essential, as it affects grain growth and, consequently, the overall performance of the materials (Milcarek & Ahn, 2020). Thus, a careful balance of dopant concentration and particle size optimization is vital for maximizing the efficiency of GDC and SDC in SOFC applications.

Fabrication Techniques for Composite Cathodes

Mixing and co-sintering processes for GDC-SDC composites

The mixing and co-sintering processes for Gadolinium-Doped Ceria (GDC) and Samarium-Doped Ceria (SDC) composites are critical for optimizing their performance in solid oxide fuel cells (SOFCs). Effective mixing techniques, such as ball milling, enhance the uniformity of the dopant distribution, which is essential for mitigating dopant segregation and improving ionic conductivity (Bae et al., 2016). Co-sintering temperatures significantly influence the microstructure and electrochemical properties of the composites. For instance, co-sintering GDC with YSZ at temperatures around 1400°C can lead to undesirable interfacial reactions that degrade ionic conductivity (Rehman et al., 2020).

The incorporation of sintering aids, such as Co, Cu, and Zn, has been shown to enhance the densification of GDC during co-sintering, allowing for lower sintering temperatures while maintaining structural integrity (Rehman et al., 2019; , Nicollet et al., 2017). Additionally, infiltration methods can be employed to fill porosity in GDC interlayers, further improving densification and conductivity (Nicollet et al., 2017; , Wang et al., 2019). The careful control of these processes is vital to achieving optimal performance in GDC-SDC composites for SOFC applications.

Thin-film deposition methods (screen printing, spray deposition, tape casting)

Thin-film deposition methods such as screen printing, spray deposition, and tape casting are pivotal in the fabrication of composite cathodes for solid oxide fuel cells (SOFCs). Screen printing is a widely used technique that allows for the precise application of cathode materials onto substrates, facilitating the creation of thick films with controlled thickness and composition. This method is particularly advantageous for large-scale production due to its simplicity and cost-effectiveness (Ke, 2024).

Spray deposition, specifically spray pyrolysis, is another effective technique for thin-film fabrication. It involves the spraying of precursor solutions onto substrates, enabling uniform coating and the ability to incorporate various dopants. This method is noted for its low-cost setup and suitability for large-area applications, making it ideal for industrial-scale production of cathodes (Sakhta, 2017). Furthermore, spray deposition can be adjusted to optimize film properties by varying parameters such as solution concentration and substrate temperature (Musa et al., 2021).

Tape casting as shown in figure 6, is a promising method that allows for the fabrication of thin-film electrodes with enhanced mechanical properties and ionic conductivity. This technique enables the production of uniform and dense films at lower temperatures, which is beneficial for integrating solid-state components in micro batteries and SOFCs (Ke, 2024). Overall, the choice of deposition method significantly impacts the performance and scalability of composite cathodes in energy applications.

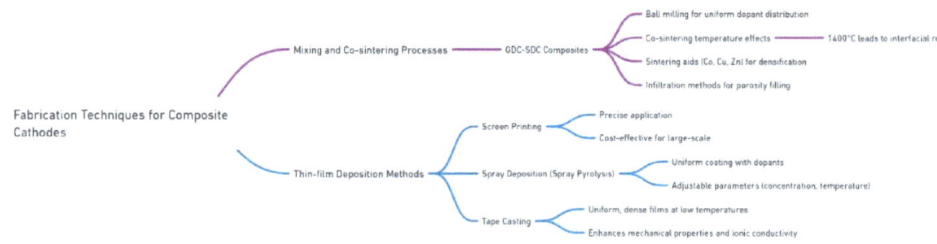

Figure 6. Fabrication Techniques for Composite Cathodes

Factors affecting microstructure and material bonding

The microstructure and material bonding of composite cathodes in solid oxide fuel cells (SOFCs) are influenced by several factors during fabrication techniques. The choice of deposition method, such as plasma spraying, screen printing, or tape casting, significantly affects the microstructure. For instance, plasma spraying can lead to variations in coating microstructure depending on the feedstock composition, which can impact the performance and stability of the cathodes (Harris & Kesler, 2011). The incorporation of secondary phases, like SDC in SSC-SDC composites, can enhance stability but may not improve initial performance, indicating a complex relationship between microstructure and functionality (Harris & Kesler, 2011).

Additionally, the volumetric ratio of components in a composite cathode plays a crucial role in determining the electrical properties and bonding characteristics. Studies have shown that the optimal ratio of LSCF to GDC in composite cathodes affects the activation energy and polarization resistance, with lower interface resistance observed at specific volumetric ratios (Muller et al., 2014). The porosity and grain size also influence the ionic conductivity and mechanical stability of the cathodes, as smaller grain sizes can enhance surface area and improve electrochemical performance (Muller et al., 2013).

Moreover, the sintering conditions, including temperature and time, are critical for achieving the desired microstructure and bonding strength. Higher sintering temperatures can promote densification but may also lead to undesirable interfacial reactions that compromise performance (Wei et al., 2010). Thus, optimizing these factors is essential for enhancing the performance and durability of composite cathodes in SOFC applications.

Challenges in Fabrication

Compatibility between GDC and SDC phases

The compatibility between Gadolinium-Doped Ceria (GDC) and Samarium-Doped Ceria (SDC) phases presents significant challenges in the fabrication of composite cathodes for solid oxide fuel cells (SOFCs). One primary concern is the potential for interfacial reactions that can occur during high-temperature sintering processes. Excessive sintering temperatures (>1300°C) can lead to the formation of resistive phases such as $(Ce,Sm)_2Zr_2O_7$, which negatively impacts ionic conductivity and overall cell performance Milcarek & Ahn (2020). This phase formation is particularly problematic at the interface between GDC, SDC, and other materials like YSZ, necessitating careful control of sintering conditions to avoid detrimental reactions (Milcarek et al., 2016).

Moreover, the lattice mismatch between GDC and SDC can influence their compatibility. The ionic radii of Gd^{3+} (1.19 Å) and Sm^{3+} (1.22 Å) are slightly different, which can lead to strain at the interface and affect the stability of the composite (Irshad et al., 2016). The optimal volumetric ratios of GDC and SDC in composites are also critical; studies suggest that achieving a balance in these ratios is essential for maximizing electronic and ionic conductivity while minimizing phase separation (Joo et al., 2014; , Lee et al., 2012).

Additionally, the microstructural characteristics, such as grain size and porosity, are influenced by the processing methods and can further affect the compatibility of the two phases. For example, finer grain sizes can enhance the three-phase boundary length, improving electrochemical performance, but may also lead to increased interfacial stress (Jiang et al., 2014). Therefore, understanding and optimizing these factors is crucial for developing effective GDC-SDC composite cathodes.

Controlling porosity, particle dispersion, and grain size

Controlling porosity, particle dispersion, and grain size is essential in the fabrication of composite cathodes for solid oxide fuel cells (SOFCs), as these factors significantly influence electrochemical performance. Porosity affects ionic conductivity and the overall electrochemical activity of the cathode. For instance, a study found that increasing the porosity of a composite cathode enhanced its performance, with optimal porosity levels around 30% facilitating better ion transport Agun et al. (2015), Dedikarni et al., 2011). However, excessive porosity can lead to mechanical instability and reduced structural integrity (Raharjo et al., 2011).

Particle dispersion is another critical factor; uniform distribution of particles within the composite enhances the connectivity between ionic and electronic conducting phases, which is vital for efficient electrochemical reactions. Research indicates that the addition of pore formers can improve particle dispersion and optimize the mass ratio of components, leading to lower polarization resistance and enhanced performance (Dong et al., 2019).

Grain size as shown in figure 7, also plays a pivotal role in determining the electrochemical properties of the cathode. Smaller grain sizes can increase the surface area available for reactions, thereby improving ionic conductivity (Muchtar et al., 2010). However, larger grains can lead to reduced porosity and hinder the oxygen reduction reaction, which is crucial for SOFC performance (Muchtar et al., 2010). Therefore, achieving a balance among these parameters through careful control of fabrication processes is essential for developing high-performance composite cathodes.

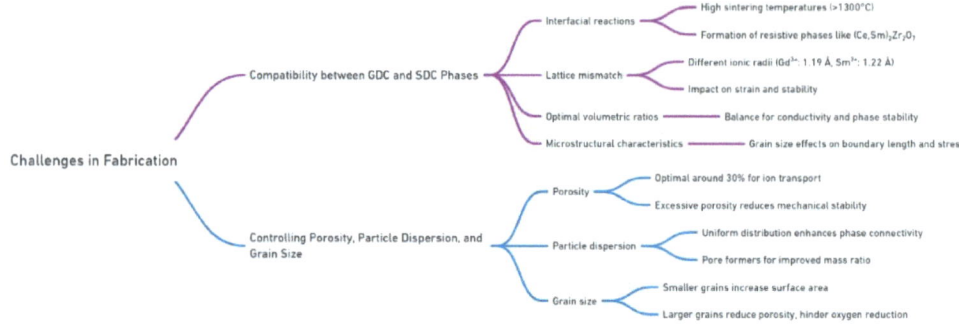

Figure 7. Challenges in Fabrication

Innovations in Synthesis and Fabrication

Novel methods for enhancing the stability and performance of GDC-SDC composites

Novel methods for enhancing the stability and performance of Gadolinium-Doped Ceria (GDC) and Samarium-Doped Ceria (SDC) composites focus on optimizing microstructural properties and improving ionic and electronic conductivity. One promising approach involves the infiltration of SDC or palladium into a scaffold of $La_{0.6}Sr_{0.4}Co_3O_3$ (LSCF), which has been reported to significantly enhance the performance and stability of the oxygen reduction reaction (ORR) in solid oxide fuel cells (SOFCs) Uchida et al. (2019). This method leverages the benefits of increased triple-phase boundary (TPB) area, which facilitates better charge and mass transfer during operation (Tayyab et al., 2021).

Additionally, the incorporation of nanostructured materials, such as Ti_3C_2Tx MXene, into SDC has demonstrated improvements in ionic conductivity due to the formation of pathways that enhance oxygen ion transport (Xian, 2019). This lamellar structure not only improves ionic conductivity but also maintains the thermal stability of the composite at elevated temperatures. Furthermore, optimizing the mass ratios of components in composite cathodes, such as LSCF and GDC, has been shown to reduce polarization resistance and improve electrochemical performance (Dong et al., 2019).

Another innovative method involves using polymeric precursors to fabricate LSCF-GDC nanocomposite thin films. This technique allows for molecular-level mixing of cations, resulting in enhanced interfacial area and improved performance due to better structural integrity and stability (Sındıraç et al., 2019). Collectively, these methods highlight the importance of material design and processing techniques in advancing the performance and durability of GDC-SDC composites for SOFC applications.

Scalable fabrication processes for commercial applications

Scalable fabrication processes for Gadolinium-Doped Ceria (GDC) and Samarium-Doped Ceria (SDC) composites are essential for commercial applications in solid oxide fuel cells (SOFCs). One promising approach is the use of inkjet printing technology, which allows for the precise deposition of materials and can be adapted for large-scale production. For instance, a modified inkjet printer has been successfully employed to fabricate LSCF-GDC composite cathodes, demonstrating the feasibility of this method for commercial applications Han et al. (2017). This technique not only enhances the uniformity of the composite but also reduces material waste compared to traditional methods.

Another scalable method is the ball milling process, which has been shown to effectively mix commercial powders of BSCF and SDC to create composite cathodes with improved performance characteristics ("Structure and Thermal

Behaviour of BSCF-SDC-Ag Composite Cathode for Solid Oxide Fuel Cell", 2019). This method allows for the production of homogeneous powders that can be easily processed into cathodes, making it suitable for industrial applications.

Additionally, the use of polymeric precursors for the fabrication of LSCF-GDC nanocomposite thin films has emerged as a cost-effective technique. This method involves molecular-level mixing of cations, resulting in enhanced interfacial area and improved electrochemical performance (Sındıraç et al., 2019). Such techniques not only streamline the fabrication process but also enhance the scalability of producing high-performance composite materials.

Overall, these innovative fabrication methods, including inkjet printing, ball milling, and polymeric precursor techniques, provide scalable solutions for the commercial production of GDC-SDC composites, addressing the challenges of uniformity, performance, and cost effectiveness in SOFC applications.

References

Ahmad, K. (2023). Preparation and characterization of nanocomposite perovskite cathode materials la0.3sr0.7fe0.4ti0.6o3-δ (lsft) for low-temperature sofcs with incorporation of graphene oxide (go). Kristall Und Technik, 59(1). https://doi.org/10.1002/crat.202300197

Feng, W., Gong, W., & Xu, Y. (2012). Lattice boltzmann simulation on solid oxide fuel cell performance. Advanced Materials Research, 472-475, 260-273. https://doi.org/10.4028/www.scientific.net/amr.472-475.260

Hajimolana, S., Hussain, M., Soroush, M., Daud, W., & Chakrabarti, M. (2012). Multilinear-model predictive control of a tubular solid oxide fuel cell system. Industrial & Engineering Chemistry Research, 121218163527007. https://doi.org/10.1021/ie301107r

Ramasamy, P. (2024). A comprehensive review on different types of fuel cell and its applications. Bulletin of Electrical Engineering and Informatics, 13(2), 774-780. https://doi.org/10.11591/eei.v13i2.6348

Singhal, S. (2000). Advances in solid oxide fuel cell technology. Solid State Ionics, 135(1-4), 305-313. https://doi.org/10.1016/s0167-2738(00)00452-5

Vázquez, S., Davyt, S., Basbus, J., Soldati, A., Amaya, A., Serquis, A., ... & Suescun, L. (2015). Synthesis and characterization of la0.6sr0.4fe0.8cu0.2o3− oxide as cathode for intermediate temperature solid oxide fuel cells. Journal of Solid State Chemistry, 228, 208-213. https://doi.org/10.1016/j.jssc.2015.04.044

Wu, X. (2023). Novel hybrid modeling and analysis method for steam reforming solid oxide fuel cell system multifault degradation fusion assessment. Acs Omega, 8(40), 36876-36892. https://doi.org/10.1021/acsomega.3c03928

Zhou, F., Jiang, Y., Gu, X., Shi, Y., & Cai, N. (2023). Direct ammonia-fed liquid metal anode solid oxide fuel cell for co-generation of hydrogen and electricity. Ecs Transactions, 111(6), 1517-1523. https://doi.org/10.1149/11106.1517ecst

Abbas, G., Raza, R., Ashfaq, M., Chaudhry, M., Khan, A., & Ahmad, I. (2013). Electrochemical study of nanostructured electrode for low-temperature solid oxide fuel cell (ltsofc). International Journal of Energy Research, 38(4), 518-523. https://doi.org/10.1002/er.3090

Ahmad, K. (2023). Preparation and characterization of nanocomposite perovskite cathode materials la0.3sr0.7fe0.4ti0.6o3−δ (lsft) for low-temperature sofcs with incorporation of graphene oxide (go). Kristall Und Technik, 59(1). https://doi.org/10.1002/crat.202300197

Bi, L., Shafi, S., Da'as, E., & Traversa, E. (2018). Tailoring the cathode–electrolyte interface with nanoparticles for boosting the solid oxide fuel cell performance of chemically stable proton-conducting electrolytes. Small, 14(32). https://doi.org/10.1002/smll.201801231

Chen, T., Pang, S., Shen, X., Jiang, X., & Wang, W. (2016). Evaluation of ba-deficient prba1−xfe2o5+δ oxides as cathode materials for intermediate-temperature solid oxide fuel cells. RSC Advances, 6(17), 13829-13836. https://doi.org/10.1039/c5ra19555a

Dong, D., Shao, X., Wang, Z., & Li, C. (2013). Cathode supports of sofcs with a hierarchical pore structure. Ecs Transactions, 57(1), 555-560. https://doi.org/10.1149/05701.0555ecst

Gunawan, G. and Setyawan, I. (2021). Progress in glass-ceramic seal for solid oxide fuel cell technology. Journal of Advanced Research in Fluid Mechanics and Thermal Sciences, 82(1), 39-50. https://doi.org/10.37934/arfmts.82.1.3950

Nawaz, H., Rafique, M., Tahir, M., Nabi, G., & Khalid, N. (2018). Material and method selection for efficient solid oxide fuel cell anode: recent advancements and reviews. International Journal of Energy Research, 43(7), 2423-2446. https://doi.org/10.1002/er.4210

Wang, T., Wang, R., Xie, X., Chang, S., Wei, T., Dong, D., ... & Wang, Z. (2022). Robust direct hydrocarbon solid oxide fuel cells with exsolved anode nanocatalysts. Acs Applied Materials & Interfaces, 14(51), 56735-56742. https://doi.org/10.1021/acsami.2c16284

Zheng, Y., Chen, X., Dong, W., & Li, J. (2016). Scaling up and characterization of single-layer fuel cells. Energy Technology, 4(8), 967-972. https://doi.org/10.1002/ente.201600001

Zhu, B., Raza, R., Abbas, G., & Singh, M. (2011). An electrolyte-free fuel cell constructed from one homogenous layer with mixed conductivity. Advanced Functional Materials, 21(13), 2465-2469. https://doi.org/10.1002/adfm.201002471

Cao, Y., Gadre, M., Ngo, A., Adler, S., & Morgan, D. (2019). Factors controlling surface oxygen exchange in oxides. Nature Communications, 10(1). https://doi.org/10.1038/s41467-019-08674-4

Chen, T., Pang, S., Shen, X., Jiang, X., & Wang, W. (2016). Evaluation of ba-deficient $prba1-xfe2o5+\delta$ oxides as cathode materials for intermediate-temperature solid oxide fuel cells. RSC Advances, 6(17), 13829-13836. https://doi.org/10.1039/c5ra19555a

Choi, H., Kim, M., Neoh, K., Jang, D., Kim, H., Shin, J., ... & Shim, J. (2016). High-performance silver cathode surface treated with scandia-stabilized zirconia nanoparticles for intermediate temperature solid oxide fuel cells. Advanced Energy Materials, 7(4). https://doi.org/10.1002/aenm.201601956

Han, H., Hu, X., Zhang, B., Zhang, S., Zhang, Y., & Xia, C. (2023). Method to determine the oxygen reduction reaction kinetics via porous dual-phase composites based on electrical conductivity relaxation. Journal of Materials Chemistry A, 11(5), 2460-2471. https://doi.org/10.1039/d2ta07293a

Nagasawa, T. and Hanamura, K. (2017). Particle-scaled visualization of active sites in lsm/scsz composite cathode of sofc through oxygen isotope labeling. Ecs Transactions, 78(1), 855-859. https://doi.org/10.1149/07801.0855ecst

Zhang, L., Huan, D., Zhu, K., Dai, P., Peng, R., & Xia, C. (2022). Tuning the phase transition of srfeo3−δ by mn toward enhanced catalytic activity and co2 resistance for the oxygen reduction reaction. Acs Applied Materials & Interfaces, 14(15), 17358-17368. https://doi.org/10.1021/acsami.2c01339

Zhou, W., Ge, L., Chen, H., Liang, F., Xu, H., Motuzas, J., ... & Zhu, Z. (2011). Amorphous iron oxide decorated 3d heterostructured electrode for highly efficient oxygen reduction. Chemistry of Materials, 23(18), 4193-4198. https://doi.org/10.1021/cm201439d

Ahmad, K. (2023). Preparation and characterization of nanocomposite perovskite cathode materials la0.3sr0.7fe0.4ti0.6o3-δ (lsft) for low-temperature sofcs with incorporation of graphene oxide (go). Kristall Und Technik, 59(1). https://doi.org/10.1002/crat.202300197

Cetin, D., Yu, Y., Luo, H., Lin, X., Ludwig, K., Basu, S., ... & Gopalan, S. (2014). Effect of carbon dioxide on the cathodic performance of solid oxide fuel cells. Ecs Transactions, 61(1), 131-137. https://doi.org/10.1149/06101.0131ecst

Jia, C., Ma, Q., Han, M., Wang, W., Menzler, N., & Guillon, O. (2019). Fabrication and performance of la, co-substituted srtio3 as cathode materials of solid oxide fuel cell. Ecs Transactions, 91(1), 1291-1298. https://doi.org/10.1149/09101.1291ecst

Jo, K., Ha, J., Ryu, J., Lee, E., & Lee, H. (2021). Dc 4-point measurement for total electrical conductivity of sofc cathode material. Applied Sciences, 11(11), 4963. https://doi.org/10.3390/app11114963

Lin, B., Chen, Y., Cheng, Z., Yang, Y., & Traversa, E. (2017). Co2-stable alkaline-earth-free solid oxide fuel cells with ni0.7co0.3o-ce0.8sm0.2o1.9composite cathodes. Ecs Transactions, 78(1), 489-497. https://doi.org/10.1149/07801.0489ecst

Agun, L., Ahmad, S., Muchtar, A., & Rahman, H. (2015). Influence of binary carbonate on the physical and chemical properties of composite cathode for low-temperature sofc. Advanced Materials Research, 1087, 177-181. https://doi.org/10.4028/www.scientific.net/amr.1087.177

Chen, D., Yang, G., Ciucci, F., Tadé, M., & Shao, Z. (2014). 3d core-shell architecture from infiltration and beneficial reactive sintering as highly efficient and thermally stable oxygen reduction electrode. Journal of Materials Chemistry A, 2(5), 1284-1293. https://doi.org/10.1039/c3ta13253f

Han, G., Choi, H., Bae, K., Choi, H., Jang, D., & Shim, J. (2017). Fabrication of lanthanum strontium cobalt ferrite-gadolinium-doped ceria composite cathodes using a low-price inkjet printer. Acs Applied Materials & Interfaces, 9(45), 39347-39356. https://doi.org/10.1021/acsami.7b11462

Lee, S. (2023). Heterogeneous composite fibrous cathode undergoing emphasized active oxygen dissociation for la(sr)ga(mg)o3-based high-performed solid oxide fuel cells. Small Structures, 5(2). https://doi.org/10.1002/sstr.202300292

Nagasawa, T. and Hanamura, K. (2017). Particle-scaled visualization of active sites in lsm/scsz composite cathode of sofc through oxygen isotope labeling. Ecs Transactions, 78(1), 855-859. https://doi.org/10.1149/07801.0855ecst

Agun, L., Ahmad, S., Muchtar, A., & Rahman, H. (2015). Influence of binary carbonate on the physical and chemical properties of composite cathode for low-temperature sofc. Advanced Materials Research, 1087, 177-181. https://doi.org/10.4028/www.scientific.net/amr.1087.177

Han, G., Choi, H., Bae, K., Choi, H., Jang, D., & Shim, J. (2017). Fabrication of lanthanum strontium cobalt ferrite-gadolinium-doped ceria composite cathodes using a low-price inkjet printer. Acs Applied Materials & Interfaces, 9(45), 39347-39356. https://doi.org/10.1021/acsami.7b11462

Ko, H., Myung, J., Lee, J., Hyun, S., & Chung, J. (2012). Synthesis and evaluation of (la0.6sr0.4)(co0.2fe0.8)o3 (lscf)-y0.08zr0.92o1.96 (ysz)-gd0.1ce0.9o2−δ (gdc) dual composite sofc cathodes for high performance and durability. International Journal of Hydrogen Energy, 37(22), 17209-17216. https://doi.org/10.1016/j.ijhydene.2012.08.099

Alemayehu, A., Zákutná, D., Kohúteková, S., & Tyrpekl, V. (2022). Transition between two solid-solutions: effective and easy way for fine ce1−xgdxo2−x/2 powders preparation. Journal of the American Ceramic Society, 105(7), 4621-4631. https://doi.org/10.1111/jace.18443

Brito, M., Izuki, M., Yamaji, K., Kishimoto, H., Shimonosono, T., Horita, T., ... & Yokokawa, H. (2011). Microstructural aspects of cation interdiffusion across the lscf/gdc interface. Ecs Transactions, 35(1), 2341-2347. https://doi.org/10.1149/1.3570230

Coppola, N., Polverino, P., Carapella, G., Sacco, C., Galdi, A., Vaiano, V., ... & Pianese, C. (2018). Structural and electrical characterization of sputter-deposited gd0.1ce0.9o2−δ thin buffer layers at the y-stabilized zirconia electrolyte interface for it-solid oxide cells. Catalysts, 8(12), 571. https://doi.org/10.3390/catal8120571

Futamura, S., Tachikawa, Y., Matsuda, J., Lyth, S., Shiratori, Y., Taniguchi, S., ... & Sasaki, K. (2017). Alternative sofc anode materials with ion- and electron-conducting backbones for higher fuel utilization. Ecs Transactions, 78(1), 1179-1187. https://doi.org/10.1149/07801.1179ecst

Li, Z., Mori, T., Ye, F., Ou, D., Zou, J., & Drennan, J. (2010). Dislocation associated incubational domain formation in lightly gadolinium-doped ceria. Microscopy and Microanalysis, 17(1), 49-53. https://doi.org/10.1017/s143192761009416x

Mulligan, J., Gopalan, S., Pal, U., & Basu, S. (2022). Quantifying the relationship between microstructure and performance in gadolinium-doped ceria infiltrated ni/ysz symmetric cells. Jom, 74(12), 4527-4532. https://doi.org/10.1007/s11837-022-05534-3

Navarrete, L., Balaguer, M., Vert, V., & Serra, J. (2016). Tailoring electrocatalytic properties of solid oxide fuel cell composite cathodes based on (la0.8sr0.2)0.95mno3+δ and doped cerias ce1−xlnxo2−δ (ln=gd, la, er, pr, tb and x=0.1−0.2). Fuel Cells, 17(1), 100-107. https://doi.org/10.1002/fuce.201600133

Ali, A., Rafique, A., Kaleemullah, M., Abbas, G., Khan, M., Ahmad, M., … & Raza, R. (2018). Effect of alkali carbonates (single, binary, and ternary) on doped ceria: a composite electrolyte for low-temperature solid oxide fuel cells. Acs Applied Materials & Interfaces, 10(1), 806-818. https://doi.org/10.1021/acsami.7b17010

Ali, M., Muchtar, A., Sulong, A., Muhamad, N., & Majlan, E. (2013). Influence of sintering temperature on the power density of samarium-doped-ceria carbonate electrolyte composites for low-temperature solid oxide fuel cells. Ceramics International, 39(5), 5813-5820. https://doi.org/10.1016/j.ceramint.2013.01.002

Deng, T., Zhang, C., Xiao, Y., Xie, A., Pang, Y., & Yang, Y. (2015). One-step synthesis of samarium-doped ceria and its co catalysis. Bulletin of Materials Science, 38(5), 1149-1154. https://doi.org/10.1007/s12034-015-0994-9

Hoa, N., Rahman, H., & Somalu, M. (2016). Preparation of nickel oxide-samarium-doped ceria carbonate composite anode powders by using high-energy ball milling for low-temperature solid oxide fuel cells. Materials Science Forum, 840, 97-102. https://doi.org/10.4028/www.scientific.net/msf.840.97

Rahmanipour, M., Pappacena, A., Boaro, M., & Donazzi, A. (2017). Distributed-charge transfer model analysis of sdc-based it-sofcs for the electrochemical oxidation of syngas and biogas. Ecs Transactions, 78(1), 1305-1318. https://doi.org/10.1149/07801.1305ecst

Yang, Q., Meng, B., Lin, Z., Zhu, X., Feng, Y., & Wu, S. (2016). Increased electrical conductivity and the mechanism of samarium-doped ceria/al2o3 nanocomposite electrolyte. Journal of the American Ceramic Society, 100(2), 686-696. https://doi.org/10.1111/jace.14471

Baharuddin, N., Muchtar, A., Somalu, M., Ali, M., & Rahman, H. (2016). Influence of sintering temperature on the polarization resistance of $la_{0.6}sr_{0.4}co_{0.2}fe_{0.8}o_{3-\delta}$ - sdc carbonate composite cathode. Ceramics - Silikaty, 115-121. https://doi.org/10.13168/cs.2016.0017

Deng, T., Zhang, C., Xiao, Y., Xie, A., Pang, Y., & Yang, Y. (2015). One-step synthesis of samarium-doped ceria and its co catalysis. Bulletin of Materials Science, 38(5), 1149-1154. https://doi.org/10.1007/s12034-015-0994-9

Irshad, M., Siraj, K., Raza, R., Javed, F., Ahsan, M., Shakir, I., ... & Rafique, M. (2016). High performance of sdc and gdc core shell type composite electrolytes using methane as a fuel for low temperature sofc. Aip Advances, 6(2). https://doi.org/10.1063/1.4941676

Kikuchi, R., Okamoto, T., Akamatsu, K., Sugawara, T., & Nakao, S. (2011). Reaction sites of mixed conductor anodes in solid oxide fuel cells. Ecs Transactions, 35(1), 1707-1715. https://doi.org/10.1149/1.3570158

Li, P., Yu, B., Li, J., Yao, X., Zhao, Y., & Li, Y. (2016). Improved activity and stability of $ni-ce_{0.8}sm_{0.2}o_{1.9}$ anode for solid oxide fuel cells fed with methanol through addition of molybdenum. Journal of Power Sources, 320, 251-256. https://doi.org/10.1016/j.jpowsour.2016.04.100

Ren, B., Li, J., Wen, G., Ricardez-Sandoval, L., & Croiset, E. (2018). First-principles based microkinetic modeling of co2 reduction at the ni/sdc cathode of a solid oxide electrolysis cell. The Journal of Physical Chemistry C, 122(37), 21151-21161. https://doi.org/10.1021/acs.jpcc.8b05312

Ali, A., Rafique, A., Kaleemullah, M., Abbas, G., Khan, M., Ahmad, M., ... & Raza, R. (2018). Effect of alkali carbonates (single, binary, and ternary) on doped ceria: a composite electrolyte for low-temperature solid oxide fuel cells. Acs Applied Materials & Interfaces, 10(1), 806-818. https://doi.org/10.1021/acsami.7b17010

Li, M., Ren, Y., Zhu, Z., Zhu, S., Chen, F., Zhang, Y., ... & Xia, C. (2016). La0.4bi0.4sr0.2feo3−δ as cobalt-free cathode for intermediate-temperature solid oxide fuel cell. Electrochimica Acta, 191, 651-660. https://doi.org/10.1016/j.electacta.2016.01.164

Liu, Y., Wang, F., Chi, B., Pu, J., Li, J., & Jiang, S. (2013). A stability study of impregnated lscf–gdc composite cathodes of solid oxide fuel cells. Journal of Alloys and Compounds, 578, 37-43. https://doi.org/10.1016/j.jallcom.2013.05.021

Milcarek, R. and Ahn, J. (2020). Micro-tubular solid oxide fuel cell polarization and impedance variation with thin porous samarium-doped ceria and gadolinium-doped ceria buffer layer thickness. Journal of Electrochemical Energy Conversion and Storage, 18(2). https://doi.org/10.1115/1.4047742

Nie, L., Liu, M., Zhang, Y., & Liu, M. (2010). La0.6sr0.4co0.2fe0.8o3−δ cathodes infiltrated with samarium-doped cerium oxide for solid oxide fuel cells. Journal of Power Sources, 195(15), 4704-4708. https://doi.org/10.1016/j.jpowsour.2010.02.049

Park, J., Yoon, H., & Wachsman, E. (2005). Fabrication and characterization of high-conductivity bilayer electrolytes for intermediate-temperature solid oxide fuel cells. Journal of the American Ceramic Society, 88(9), 2402-2408. https://doi.org/10.1111/j.1551-2916.2005.00475.x

Rahmanipour, M., Pappacena, A., Boaro, M., & Donazzi, A. (2017). A distributed charge transfer model for it-sofcs based on ceria electrolytes. Journal of the Electrochemical Society, 164(12), F1249-F1264. https://doi.org/10.1149/2.1911712jes

Sutradhar, N., Sinhamahapatra, A., Pahari, S., Jayachandran, M., Subramanian, B., Bajaj, H., ... & Panda, A. (2011). Facile low-temperature synthesis of ceria and samarium-doped ceria nanoparticles and catalytic allylic oxidation of cyclohexene. The Journal of Physical Chemistry C, 115(15), 7628-7637. https://doi.org/10.1021/jp200645q

Takamura, H., Nemoto, N., Yamaguchi, M., Kobayashi, K., Oikawa, I., Takano, A., ... & Takamura, H. (2023). Key role of interfacial cobalt segregation in stable low-resistance composite oxygen-reducing electrodes. Acs Applied Materials & Interfaces, 15(29), 34809-34817. https://doi.org/10.1021/acsami.3c04940

Yue, X. and Irvine, J. (2012). Impedance studies on lscm/gdc composite cathode for high temperature co2 electrolysis. Ecs Transactions, 41(33), 87-95. https://doi.org/10.1149/1.3702415

Zhang, L., Huang, J., Song, Z., Fu, Y., Liu, M., & He, T. (2014). Evaluation and optimization of $ba_{0.2}sr_{0.8}co_{0.9}nb_{0.1}o_{3-\delta}$-$gd_{0.1}ce_{0.9}o_{1.95}$ composite cathodes for it-sofcs. Materials Science Forum, 787, 221-226. https://doi.org/10.4028/www.scientific.net/msf.787.221

Zhang, X., Robertson, M., Decès-Petit, C., & Kesler, O. (2009). Composite cathode study for low temperature sofc. Ecs

Transactions, 25(2), 2463-2471. https://doi.org/10.1149/1.3205801

Min, Z. and Liu, X. (2013). Electrical properties of la$_{0.7}$sr$_{0.3}$cuo$_{3-δ}$cathode based on sdc-ysz composite electrolyte. Advanced Materials Research, 702, 224-228. https://doi.org/10.4028/www.scientific.net/amr.702.224

Mishina, T., Fujiwara, N., Tada, S., Takagaki, A., Kikuchi, R., & Oyama, S. (2020). Calcium-modified ni-sdc anodes in solid oxide fuel cells for direct dry reforming of methane. Journal of the Electrochemical Society, 167(13), 134512. https://doi.org/10.1149/1945-7111/abba65

Myoujin, K., Ichiboshi, H., Kodera, T., & Ogihara, T. (2011). Characterization of samarium doped ceria powders having high specific surface area synthesized by carbon-assisted spray pyrolysis. Key Engineering Materials, 485, 137-140. https://doi.org/10.4028/www.scientific.net/kem.485.137

Yang, C., Cheng, J., He, H., & Gao, J. (2010). Ni/sdc materials for solid oxide fuel cell anode applications by the glycine-nitrate method. Key Engineering Materials, 434-435, 731-734. https://doi.org/10.4028/www.scientific.net/kem.434-435.731

Zhang, Z., Zhou, W., Chen, Y., Chen, D., Chen, J., Liu, S., ... & Shao, Z. (2015). Novel approach for developing dual-phase ceramic membranes for oxygen separation through beneficial phase reaction. Acs Applied Materials & Interfaces, 7(41), 22918-22926. https://doi.org/10.1021/acsami.5b05812

Ahmad, S., Bakar, M., Muchtar, A., Muhamad, N., & Rahman, H. (2012). The effect of milling speed and calcination temperature towards composite cathode lscf-sdc carbonate. Advanced Materials Research, 576, 220-223. https://doi.org/10.4028/www.scientific.net/amr.576.220

Fu, Y., Li, C., & Hu, S. (2012). Comparison of the electrochemical properties of infiltrated and functionally gradient sm0.5sr0.5coo3−δ−ce0.8sm0.2o1.9composite cathodes for solid oxide fuel cells. Journal of the Electrochemical Society, 159(5), B629-B634. https://doi.org/10.1149/2.017206jes

Milcarek, R. and Ahn, J. (2020). Micro-tubular solid oxide fuel cell polarization and impedance variation with thin porous samarium-doped ceria and gadolinium-doped ceria buffer layer thickness. Journal of Electrochemical Energy Conversion and Storage, 18(2). https://doi.org/10.1115/1.4047742

Bae, J., Lim, Y., Park, J., Lee, D., Hong, S., An, J., ... & Kim, Y. (2016). Thermally-induced dopant segregation effects on the space charge layer and ionic conductivity of nanocrystalline gadolinia-doped ceria. Journal of the Electrochemical Society, 163(8), F919-F926. https://doi.org/10.1149/2.1201608jes

Nicollet, C., Waxin, J., Dupeyron, T., Flura, A., Heintz, J., Ouweltjes, J., ... & Bassat, J. (2017). Gadolinium doped ceria interlayers for solid oxide fuel cells cathodes: enhanced reactivity with sintering aids (li, cu, zn), and improved densification by infiltration. Journal of Power Sources, 372, 157-165. https://doi.org/10.1016/j.jpowsour.2017.10.064

Rehman, S., Shaur, A., Kim, H., Joh, D., Song, R., Lim, T., ... & Lee, S. (2020). Effect of transition metal doping on the sintering and electrochemical properties of gdc buffer layer in sofcs. International Journal of Applied Ceramic Technology, 18(2), 511-524. https://doi.org/10.1111/ijac.13650

Rehman, S., Shaur, A., Song, R., Park, S., Lim, T., Hong, J., ... & Lee, S. (2019). Enhancing the sinterability of gadolinium-doped ceria by wet chemical processing. Ecs Transactions, 91(1), 1201-1207. https://doi.org/10.1149/09101.1201ecst

Wang, G., Jia, C., Sun, Z., Chen, M., & Han, M. (2019). In situ densification of gadolinium-doped ceria interlayer by infiltration process in sofc. Ecs Transactions, 91(1), 1149-1156. https://doi.org/10.1149/09101.1149ecst

Ke, B. (2024). Tape-casting electrode architecture permits low-temperature manufacturing of all-solid-state thin-film microbatteries. Interdisciplinary Materials, 3(4), 621-631. https://doi.org/10.1002/idm2.12174

Musa, A., Ahmed, A., Said, M., Tsoho, M., & Suleiman, A. (2021). Thin films growth of sno_2:f/cds/cdte, and studies of their physical and optical properties using spray pyrolysis techniques. Asian Journal of Research and Reviews in Physics, 19-31. https://doi.org/10.9734/ajr2p/2021/v4i430149

Sakhta, A. (2017). Morphological and optical properties of pure and mg doped tin oxide thin films prepared by spray pyrolysis method. American Journal of Nanosciences, 3(2), 19. https://doi.org/10.11648/j.ajn.20170302.11

Harris, J. and Kesler, O. (2011). Performance of metal-supported composite and single-phase cathodes based on lscf and ssc. Ecs Transactions, 35(1), 1927-1934. https://doi.org/10.1149/1.3570182

Muller, G., Ringuedé, A., & Laberty-Robert, C. (2014). Synthesis, characterization and electrical properties of $la0.7sr0.3co0.2fe0.8o3/gd-ceo2$ thin films (≤ 500 nm). Journal of Materials Chemistry A, 2(18), 6448. https://doi.org/10.1039/c3ta14505k

Muller, G., Vannier, R., Ringuedé, A., Laberty-Robert, C., & Sanchez, C. (2013). Nanocrystalline, mesoporous nio/ce0.9gd0.1o2−δ thin films with tuned microstructures and electrical properties: in situ characterization of electrical responses during the reduction of nio. Journal of Materials Chemistry A, 1(36), 10753. https://doi.org/10.1039/c3ta11175j

Wei, B., Lü, Z., Wei, T., Jia, D., Huang, X., & Su, W. (2010). Enhanced performance of the gdbaco2o5+δ cathode with active ce0.8sm0.2o1.9 nanoparticles. Ecs Transactions, 28(11), 227-233. https://doi.org/10.1149/1.3495845

Irshad, M., Siraj, K., Raza, R., Javed, F., Ahsan, M., Shakir, I., ... & Rafique, M. (2016). High performance of sdc and gdc core shell type composite electrolytes using methane as a fuel for low temperature sofc. Aip Advances, 6(2). https://doi.org/10.1063/1.4941676

Jiang, W., Wei, B., Lü, Z., Wang, Z., Zhu, X., & Zhu, L. (2014). Co-synthesis of sm0.5sr0.5coo3-sm0.2ce0.8o1.9 composite cathode with enhanced electrochemical property for intermediate temperature sofcs. Fuel Cells, 14(6), 966-972. https://doi.org/10.1002/fuce.201400022

Joo, J., Yun, K., Lee, Y., Jung, J., Yoo, C., & Yu, J. (2014). Dramatically enhanced oxygen fluxes in fluorite-rich dual-phase membrane by surface modification. Chemistry of Materials, 26(15), 4387-4394. https://doi.org/10.1021/cm501240f

Lee, J., Vito, N., Yoon, H., & Wachsman, E. (2012). Effect of ni-gd0.1ce0.9o1.95anode functional layer composition on performance of lower temperature sofcs. Journal of the Electrochemical Society, 159(7), F187-F193. https://doi.org/10.1149/2.009207jes

Milcarek, R. and Ahn, J. (2020). Micro-tubular solid oxide fuel cell polarization and impedance variation with thin porous samarium-doped ceria and gadolinium-doped ceria buffer layer thickness. Journal of Electrochemical Energy Conversion and Storage, 18(2). https://doi.org/10.1115/1.4047742

Milcarek, R., Wang, K., Falkenstein-Smith, R., & Ahn, J. (2016). Performance variation with sdc buffer layer thickness. International Journal of Hydrogen Energy, 41(22), 9500-9506. https://doi.org/10.1016/j.ijhydene.2016.04.113

Agun, L., Ahmad, S., Muchtar, A., & Rahman, H. (2015). Influence of binary carbonate on the physical and chemical properties of composite cathode for low-temperature sofc. Advanced Materials Research, 1087, 177-181. https://doi.org/10.4028/www.scientific.net/amr.1087.177

Dong, L., Liu, C., Su, D., Dai, C., & Xiong, Y. (2019). Effect of microstructure on electrochemical performance of nano-structured $la_{0.8}sr_{0.2}co_{0.2}fe_{0.8}o_3-\delta-gd_{0.2}ce_{0.8}o_{1.9}$ composite cathodes. Ecs Transactions, 91(1), 1483-1489. https://doi.org/10.1149/09101.1483ecst

Muchtar, A., Hamid, N., Muhamad, N., & Daud, W. (2010). Sintering effects on lscf cathodes for intermediate temperature solid oxide fuel cells (it-sofcs). Advanced Materials Research, 139-141, 141-144. https://doi.org/10.4028/www.scientific.net/amr.139-141.141

Muchtar, A., Muhamad, N., & Daud, W. (2011). Microstructure characterisation of $ag_2o_3-bi_2o_3$ composite cathodes for intermediate temperature solid oxide fuel cells (it-sofcs). Key Engineering Materials, 471-

472, 97-102. https://doi.org/10.4028/www.scientific.net/kem.471-472.97

Raharjo, J., Muchtar, A., Daud, W., Muhamad, N., & Majlan, E. (2011). Fabrication of porous lscf-sdc carbonates composite cathode for solid oxide fuel cell (sofc) applications. Key Engineering Materials, 471-472, 179-184. https://doi.org/10.4028/www.scientific.net/kem.471-472.179

Dong, L., Liu, C., Su, D., Dai, C., & Xiong, Y. (2019). Effect of microstructure on electrochemical performance of nano-structured la0.8sr0.2co0.2fe0.8o3-δ-gd0.2ce0.8o1.9 composite cathodes. Ecs Transactions, 91(1), 1483-1489. https://doi.org/10.1149/09101.1483ecst

Sındıraç, C., Ahsen, A., Öztürk, O., Akkurt, S., Birss, V., & Büyükaksoy, A. (2019). Fabrication of lscf and lscf-gdc nanocomposite thin films using polymeric precursors. Ionics, 26(2), 913-925. https://doi.org/10.1007/s11581-019-03262-4

Tayyab, Z., Rauf, S., Chen, X., Wang, B., Shah, M., Mushtaq, N., … & Asghar, M. (2021). Advanced lt-sofc based on reconstruction of the energy band structure of the lini0.8co0.15al0.05o2-sm0.2ce0.8o2-δ heterostructure for fast ionic transport. Acs Applied Energy Materials, 4(9), 8922-8932. https://doi.org/10.1021/acsaem.1c01186

Uchida, H., Nishino, H., Kakinuma, K., & Brito, M. (2019). Further improvement of performances and durability of oxygen and hydrogen electrodes for reversible solid oxide cells. Ecs Transactions, 91(1), 2379-2386. https://doi.org/10.1149/09101.2379ecst

Xian, H. (2019). Effect of mxene on oxygen ion conductivity of sm0.2ce0.8o1.9 as electrolyte for low temperature sofc. International Journal of Electrochemical Science, 14(8), 7729-7736. https://doi.org/10.20964/2019.08.07

(2019). Structure and thermal behaviour of bscf-sdc-ag composite cathode for solid oxide fuel cell. International Journal of Engineering and Advanced Technology, 9(2), 1582-1585. https://doi.org/10.35940/ijeat.b2371.129219

Han, G., Choi, H., Bae, K., Choi, H., Jang, D., & Shim, J. (2017). Fabrication of lanthanum strontium cobalt ferrite-gadolinium-doped ceria composite cathodes using a low-price inkjet printer. Acs Applied Materials & Interfaces, 9(45), 39347-39356. https://doi.org/10.1021/acsami.7b11462

Sındıraç, C., Ahsen, A., Öztürk, O., Akkurt, S., Birss, V., & Büyükaksoy, A. (2019). Fabrication of lscf and lscf-gdc nanocomposite thin films using polymeric precursors. Ionics, 26(2), 913-925. https://doi.org/10.1007/s11581-019-03262-4

www.ingramcontent.com/pod-product-compliance
Lightning Source LLC
Chambersburg PA
CBHW040249220526
45473CB00001B/418